Think and Become Safety Practitioner

Think and Become Safety Practitioner

**JESUS G. PEDINES JR
OSHP**
1st Edition 2015

Copyright © 2015 by Jesus G. Pedines Jr.

Library of Congress Control Number: 2015901825
ISBN: Hardcover   978-1-5035-0235-2
               Softcover   978-1-5035-0234-5
               eBook   978-1-5035-0233-8

All rights reserved. No part of this book may be reproduced or transmitted in any form or by any means, electronic or mechanical, including photocopying, recording, or by any information storage and retrieval system, without permission in writing from the copyright owner.

Any people depicted in stock imagery provided by Thinkstock are models, and such images are being used for illustrative purposes only.
Certain stock imagery © Thinkstock.

Print information available on the last page.

Rev. date: 02/24/2015

To order additional copies of this book, contact:
Xlibris
1-800-455-039
www.Xlibris.com.au
Orders@Xlibris.com.au
704376

CONTENTS

Introduction 7
A Brief Background About The Author 11

Lesson 1:   Pray For It 15
Lesson 2:   Why The Safety Profession? 18
Lesson 3:   My Auntie's Smile Is 50 Per Cent Wider 21
Lesson 4:   Our Parents Are Safety Practitioners 23
Lesson 5:   Improved Writing And Reading Skills At Home 26
Lesson 6:   The Safety Profession Is My Life Now 28
Lesson 7:   The Importance Of Being A Member
            Of A Safety Organization Or Society 31
Lesson 8:   Personal KPI Checklist 34
Lesson 9:   How To Approach Safety Violators In The Workplace ... 37
Lesson 10: Sample Photos Of Basic Unsafe Acts,
            Unsafe Conditions 42
Lesson 11: Recommended Safety Training
            For New Safety Officers 51
Lesson 12: Messages From My Mentors 54
Lesson 13: A Reminder When You Get A Job 59
Lesson 14: Unsafe Act And Unsafe Condition Report Exercises 64

A Message of Thanks from the Author 143

Introduction

AssalamuAlaikum. May the peace and blessing of our Almighty Creator be upon you.

As part of this handbook, a survey was conducted for at least 150 individuals (Filipino, Indian, Bangladesh, Egyptian, and some Saudi nationals and several colleagues from western countries), where 70 per cent among them were practicing safety and the other 30 per cent were those individuals willing or planning to be safety practitioners someday. From the 70 per cent safety practitioners, about 99 per cent of those individuals who participated in the survey mentioned that they were happy to be a safety practitioner and looking forward to uplift their credentials, experience, skills, and knowledge to be a successful safety professional in the future. And those 30 per cent are on their way, preparing themselves, attending safety trainings and seminars, and planning to shift from their current jobs to a safety profession.

My objective is to share my journey as a safety practitioner. I hope you will find beneficial information to guide you based on my twelve years' personal experience in the construction industry as a safety professional. For those who are willing to be a safety practitioner, male or female, having a bachelor degree or a college undergraduate, as long as you are willing to learn and want to be a member of a safety society and be part of the team in saving lives in a workplace around the globe—it could be as a safety officer, safety inspector, supervisor, advisor, analyst, auditor, environmentalist, or manager and so on—then I would like to welcome you all.

Thanks and praise to Almighty God. My desire and love of my profession as a safety practitioner inspired me to write this book, so I told myself why not share my journey and advocacy of being safety practitioner to those who are willing to try this field and to all my co–safety professionals who wish to level up their current designation? Saving human life is wonderful wisdom and the duty of an individual. In Islam, we believe that if a person removes anything that could hurt someone on the street or walkways, he/she will get a reward from Almighty Creator, when in fact this is one of the basic duties of a safety officer in the workplace. Also, we believe that if anyone saves a single human life, it is as if he/she saves all humanity.

Why are safety professionals very important in an organization? And why are safety personnel needed in the construction industries and other general industries to maintain and promote safety, health, and environment and protect property in the workplace all over the world? Why? Well, it's because of moral, economic, and legal reasons and obligations. On 2012 as per International Labour Organization (ILO[1]), statistics show that there are 270 million occupational accidents and 160 million occupational diseases recorded each year. two million die every year from occupational accidents and diseases, and 4 per cent of world gross domestic products are lost each year.

Is there any shortage of safety practitioners in the workplace? Yes, there is. Very few individuals who are currently working as safety professionals and the competition rate in applying for this position is not that high compared to other positions that most applicants apply for, such as office staff, secretaries, engineers, nurses, accountants, technicians,

---

[1] ILO statistics report quoted from author NEBOSH IGC-1 training material March 2012 Edition by RRC Training through training provider 966 - Jabal Dami Industrial Services & Technical Support, Saudi Arabia.

operators, factory workers, and so on. Safety practitioner applicants are very few in the market—that's why your chances of being hired are higher than other positions. So don't stop, keep going, don't entertain negative thoughts like "I can't do it, I am not qualified" and so on. Don't be afraid of those negative thinkers who are always criticizing people; instead use that as a challenge or tool to conquer your fear of being criticized by other people. Just stay away from those kinds of people and find a friend who is always a positive thinker and has a proactive mindset.

This pocket-sized book may help you to figure out the know-how to be a qualified safety practitioner regardless of your nationality, faith, and current educational achievements.

Welcome

A BRIEF BACKGROUND ABOUT THE AUTHOR

During my childhood, I believed that I could do anything, especially if someone pushed me to do things, even if I knew they were dangerous. My loving mother knew my attitude very well—that's why she did her best to discipline me, but still I didn't listen that much to her advice and warnings. The consequences were that a stick was always waiting for me when I got home. In my opinion, my skills and talents during my early age, when I was five to ten years old were unbelievable for an ordinary kid. I was a fisherman, a tree climber, a sales kid, a singer and guitarist, a karate kid, and a cook. I was good in acting and dancing, and my favorite was being a team leader; my friends always wanted me to be their leader. Until now, I still love to use some of those talents and skills I possess. Thanks to the almighty Creator.

During my high school and college, I had shown the same talents to my classmates, friends, and family. I worked for a living to help my parents. During high school, every summer vacation, I worked in the university orange plantation, and I sold newspapers during weekends. During college, I sold fried peanuts to my schoolmates in the library. I worked in a fast-food restaurant as service crew, joined a dance group, composed a song, joined a rock band, and competed in various singing contests to earn some cash prices. I tried modeling and eventually joined a male pageant (Ginoong Pilipinas 1997), representing the province of Nueva Vizcaya, Philippines. Every vacation I loved to go fishing, and I cooked for my family.

My leadership skills were developed when I and four other friends founded a youth organization in our village called Volunteer Brigade. Our goal and objective was to provide free services to our community. We helped community leaders on their clean-and-green programs such as tree planting along the roads. At school, I joined Citizenship Advancement Training (CAT) and became an officer, a platoon leader. During college, I joined several collegiate organizations and fraternities such as Alpha Phi Omega, an international service fraternity and sorority. I survived in 1995 with ID # 23012 at Theta Epsilon Chapter and survived the Reserve Officers' Training Corps (ROTC) and became company commander handling Bravo Company at Nueva Vizcaya State University. I almost tried most of the extracurricular activities during those days, but I never failed any subjects, although I had two incomplete subjects and no grade and one for remedial, but I was able to manage them, take the retest, and submit my projects on time, and I passed those subjects. My goal during college was not to fail any subject because I had limited financial resources, and my parents could not afford if I would repeat any subjects, so I did my best and kept my desire to pass all my subjects.

My father was a carpenter. He worked abroad for seven years and retired from being an overseas Filipino worker (OFW) when I was in first year in high school. My father bought secondhand a very old tricycle for public transportation to make a living while my loving mother accepted laundry jobs to help my father to provide for the family's basic needs. Our parents' priority was to provide for our basic needs in school and for living. Their hardship in running our family was one of the reasons and the inspiration behind why my eldest sister, my younger brother, and I needed to be good in school to be able to finish our bachelor degrees. Education was the only thing they could give us to not be like them, they said. So this is who I am, having grown up in a poor but happy family living in a small house without windows. Thanks and praise to Almighty God, we made them proud because we did it. My sister graduated Bachelor of Arts, major in English, at Nueva Vizcaya State University, Philippines; our younger brother graduated Bachelor of Science in Civil Engineering at TIP, Manila, Philippines; and I graduated Bachelor of Science in Electrical Engineering at Manuel L. Quezon University, Philippines. Thanks to Almighty God.

Anyway, let's get started. The moment you decided to purchase a copy of this book, congratulations, you possessed a desire to shift from your current job to the safety profession. This book is really for you, whatever your race and faith. However, if you are just curious about it, I promise you'll get something beneficial once you finish reading this book, whatever your plan is. For those who really want to be a safety practitioner, just follow instructions and try your best to finish all exercises. So are you ready? Say it aloud: 'Yes, I am ready.'

Lesson 1

PRAY FOR IT

Yes, pray for it, and keep the burning desire inside you. If you are praying five times a day, then pray for your plan five times a day or more. Letting go of your tears while praising and begging our almighty Creator to guide you to fulfil your plan is a sign of having strong faith and being humble when communicating to our most merciful Creator. God is always listening to our prayers, especially when we are sincere and emotionally communicating to our most merciful Creator. The power of prayer is a miracle. It makes our faith stronger, it gives us peace of mind, it develops more love in our heart, it creates tons of happiness inside and outside us, it strengthens our hope and determination to face challenges in our life, and it can turn us into individuals with wonderful wisdom so that everyone around us will be benefited. The other way around, prayers can relieve our fears and sadness, remove jealousy and hatred, and delete anger, greed, and revenge inside our hearts.

Believe on the power of prayers, and maintain that burning desire inside you. The power of words is also unbelievable, so *write down* a statement about *who* you want to become someday. Make it a habit to read your statement aloud *positively* at least two times a day, like 'By next year, I will be working as a safety officer' or 'A year from now, I will be working abroad as a safety practitioner.' Doing this exercise every day will keep your brain working twenty-four hours a day, and you may create ideas and plan on how to achieve your goal. Once related ideas pop up in your mind, write them down quickly on a sheet of paper so you don't miss them.

Now list down all the reasons why you want to work as a safety practitioner. The space below is provided for your use.

My reasons are . . . (Use an extra sheet of paper if needed.)
_____
_____
_____
_____

Always remember your reasons. Those reasons you listed will be your inspiration to attain your goal, so include them in your prayers. Sometimes, opportunity is already in front of you but you fail or ignore those opportunities or you may be able to notice but you can't do anything because you are not ready or you have lack of self-confidence—that's why you pass that opportunity even though you'd have loved to try.

Your *desire* to be a safety practitioner will be like your transportation or vehicle to attain your goals, and your *reasons* will be like your gasoline or like your fully-charged battery to push you and keep you going to reach your destination successfully. Talk to your family or friends, and discuss with them about your plan. Ask for their support. If you are currently working but don't have sufficient extra cash to fund your plan, then your family should understand that you need their help, and you need to find ways to save and invest some funds for this goal. If you are currently jobless, then ask your family or friends who can lend you some financial assistance and promise them to pay them back as soon as you get your job. To justify why you are asking their support, just tell them your reasons. And don't forget to extend your thanks and appreciation once they agree on your plan.

For those who haven't graduated college or are just high school graduates and don't have any experience in the safety profession, just maintain your desire and finish reading this book. For those who are currently working as safety practitioners or working in other fields, reading this book may provide you some information that may help you to level up your current position, mindset, and personal development. You might be more qualified in terms of credentials and skills, getting a higher pay cheque than me, but since you bought a copy of this handbook, please finish reading till the last lesson if you think you can learn something beneficial to your plan in life, or you might recommend this book to your friends or relatives who are interested to be safety professionals.

Do not say 'It's impossible' because nothing is impossible; just pray for it. If you've failed once, don't stop; try again, again, and again as Thomas Edison didn't stop until he was able to succeed in what he wanted. He said, 'I have not failed. I've just found ten thousand ways that won't work.' In the safety profession, as long as you are willing to invest your time and invest some funds for your training, and you can at least read and write and are physically fit to work, your *burning desire* will push you to keep you going towards your decided destination. Qualifications, credentials, and required skills to be a safety practitioner can be acquired through training and regular hands-on practice.

Lesson 2

WHY THE SAFETY PROFESSION?

Once again, do not stop learning; keep going. Here are some practical reasons to further explain why I am encouraging you to become a safety practitioner. With all due respect, let's say you are currently working in a non–white collar job and are not earning sufficient salary compared to those people working in white collar jobs. If you have some safety training certificates, for sure you are one step ahead from those applicants who don't have safety training even though they might be degree holders. In the first place, they will not apply for safety positions unless they would love to try or they are aware of the requirements and they are interested to undergo safety training as you are.

So even you don't have a college diploma or degree, if you've attended some safety training and understand the characteristics of being a safety practitioner even though you don't have any experience, you have a greater chance of being hired than those applicants who never attended safety training. Once you apply for a safety officer job, it can be difficult to find a company offering jobs to inexperienced candidates because most of the companies are looking for experienced candidates and those who possess safety credentials, but as I said, don't forget the power of *prayers*, so just do it—keep your burning desire to continue learning, and you will be qualified soon, and Almighty God may give you the job you pray for.

Now if you are working in any non–white collar job or in any other fields or are currently jobless and you are happy with your current situation, then continue what you are doing. It is your choice.

But someday, if you wish and decide to work in or shift to a safety profession job, then finish reading this book.

See the table below and compare the salary grade.

Monthly Salary Grade in Philippine Pesos for Filipino Citizens (example: Those working abroad in Saudi Arabia, 2013)	
Employed as Technician	Employed as New Safety Officer
Monthly income < P17,655.00 up to P23,540.00	Monthly income > P29,425.00 up to P41,195.00

Did you notice the difference between the monthly paychecks on the table above? In a few years, you could get a higher salary as you level up your position. Most of the time safety positions get more overtime work than other employees in the job site because most of the time, safety personnel are the first to be present and shall leave the job site last, so overtime work is an extra plus for safety practitioner and one way to speed up monthly income, isn't it? However, having said that, you have to learn to be smart in spending your money; remember the difference between 'what I need' and 'what I want'. Learn how to manage earnings while working abroad and the difference between what we *need* and what we *want*. Factors such as high demand of the family needs and cost of living is one of many reasons why 70 per cent of OFWs are facing difficulties in saving money. A good example is if an individual gets a higher paycheck, his family also increases their expenses. I hope you receive my message and become a successful OFW.

Be reminded also that a safety officer's duties and responsibilities are challenging and not an easy job to those beginners because you will be one of those in the front line monitoring the safety rules' implementation in the job site. Every day you're tasked to conduct inspections and identify potential hazards that could lead to incidents or personal injuries and property damage if not corrected in a timely manner. Your recommendation to resolve your findings and to minimize or eliminate the identified hazard is very important. Furthermore, make sure that your concern has been addressed to the right person for their immediate corrective action. More details on safety practitioner characteristics, duties, and responsibilities will be mentioned on lesson 6. Don't worry, you will be equipped with those capabilities, skills, and knowledge because as I said, just keep going; do not stop learning. Such skills and qualifications can be acquired by attending safety trainings and seminars and practicing it regularly. Just like in martial arts, an artist might not be able to perform excellent kata if he/she fails to practice on a regular basis.

Just maintain your desire, and be passionate in what you are doing, and always think positive, and don't forget lesson 1.

Exercise 2: Answer the personal question below.

Present Job vs. Future Job	
What is your current position or job?	
What is your current monthly paycheck or salary?	
What is the future position you desire to have?	
What is the monthly paycheck you want to receive in the future?	

The details you have written in the above exercise will help keep your burning desire alive, help you to aim for your target, and help you to focus on your chosen direction.

Lesson 3

MY AUNTIE'S SMILE IS 50 PER CENT WIDER

I used to work with my uncle Lope Hubalde with the same company in the Philippines when I was a safety officer twelve years ago. My uncle was working as an electrician within twenty years or more from different companies, and then one day, both of us got opportunities to work abroad; he was assigned in Qatar while I was hired to work as a fire safety officer for Al-Mana General Hospital, Al-Hasa branch, Saudi Arabia. After my uncle's contract completion in Qatar, he reapplied to work abroad but this time in Saudi Arabia. After working seven years in the Kingdom as an electrician, one day I encouraged him to attend safety training and join our society, the Philippines Society of Safety Practitioners–Middle East Region.

I told my uncle to talk to my auntie and discuss my suggestion and set a plan on how they could manage a monthly budget and spare some cash for his training expenses. I still remember that I sponsored him on his first training in the society. He did not regularly attend the monthly seminar due to financial shortage, but whenever he had extra budget, he attended training. A year later, he was able to obtain around ten to fifteen safety training certificates in different topics and eventually developed his guts to apply as a safety officer. Guess what?

He is currently working as safety supervisor in the Kingdom, and the good news is that he was able to start developing his skills in the field as a safety practitioner and is now earning double his last salary grade. As a result, there 50 per cent level-up smile on both my uncle and aunt after a year of striving and saving some money for his training.

Another short testimony that I want to share with you is the story of Jonel N. Lazatin an ordinary office tea-boy working in Saudi Arabia. He underwent challenges and eventually became a safety officer, earning a 300 per cent paycheck increase. Riding in a trailer truck loaded with sulfur chemicals just to attend training was not an easy decision to make, especially if an individual is aware of the potential hazard of this chemical. This is one amazing testimony in our society that inspired many. He said, 'My dedication, determination and hard work in order to achieve my dream for my family resulted in a lot of help for my family.'

There are many successful testimonies in our society as well as in other safety organizations around the globe for sure. Some of them came from being technicians, secretaries, helpers, heavy-equipment operators, service crew, accountants, and engineers and are now working as safety practitioners, and someday, hopefully, you will be the next. If God wills.

Being a qualified safety practitioner is not easy for those who are not willing to learn because continued learning and training attendance is necessary in this field to obtain adequate knowledge. In fact, you can gain a lot of information using your fingertips online—Google it, use the technology; it is just a matter of being familiar. If you are not familiar, ask someone to teach you how to use the computer and Internet so you can benefit from the advantage of technology and can spend some of your time reading and watching videos about safety and practice what you've learned from it and level up the smile of your loved ones and be proud someday once you become a safety practitioner.

Lesson 4

OUR PARENTS ARE SAFETY PRACTITIONERS

Thirty-two years ago, during my childhood, I still remember how my loving mother and father never stop warning us about those dangerous kid habits like climbing trees, going in the forest, swimming in the river without their presence, playing fire and cooking games. Our parents advised us to avoid fighting with other kids and going to neighbors' houses that had dangerous animals like dogs. At home, they prepared food and fed us healthy food like vegetables such as sweet potato leaves, *talbos ng kamote* in the Tagalog language. Mmmmmm, I love that food. Our parents taught us how to maintain personal hygiene and made sure we had regular sleep at noontime and got to bed on time in the evening so we would grow faster and healthier, they said. If we were sick or injured, they were worried, so immediately they had to do some traditional first aid or bring us to the hospital if necessary. For our safety, my mother secured hot or boiling water, wiped off the slippery surfaces, and kept knives and household chemicals safely where we couldn't reach.

Well, maybe being a naughty and adventurous kid I ignored hazards around me because I didn't know about them. I felt cool during my childhood; that's why I did not listen to the warnings of my parents till I would get hurt and learn from it. Imagine, I fell from fruit-bearing trees seven times and almost died, experienced dog bites several times, and got minor injuries due to slip, trip, and fall incidents. For sure all parents do the same, and if you are parents now like me, I believe you are doing the same thing to keep your kids safely at home because that's a basic parenthood duty and responsibility with regard to the safety of our kids. Our parents are safety officers at home, aren't they? No parents want to see their children being hurt, am I right?

I personally advise you to love your parents more than anyone else. Let them feel how much you love them. All parents always pray for their children's safety and success in their life. So please take good care of them, financially and emotionally, especially your mother who cared extremely for you for nine months in her tummy before you were born. If they have already passed away, it's okay to cry if you feel sorry for not being kind and supportive, and weren't able to let them feel that you loved them so much when they were still alive.

Now stop for a while, wipe off your tears, and recall your childhood moments of being naughty to make you laugh this time.

Okay, let's move on. For the sake of learning, once you get home, go around your house and check your backyard, storeroom, kitchen, toilet corridors, garage, and bedrooms and try to identify some potential hazard that may hurt someone, especially your kids. If you are currently working in a construction site or manufacturing or other industries, try to inspect your workplace. Use the blank space below.

a. Potential hazard / dangerous situation (Use an extra sheet of paper if necessary.)

1. _____
2. _____
3. _____
4. _____
5. _____

Now what is your recommendation to rectify or minimize or eliminate the dangerous or unsafe conditions? List below.

b. Control measures to resolve the problem (Use an extra sheet of paper if necessary.)

1. _____
2. _____
3. _____
4. _____
5. _____

So how is your home or workplace? Is it a safe condition for everyone?

The above exercises might be easy for those who are currently working as safety practitioners. But for those who are not in safety field yet, it might be difficult, and it might take several minutes to finish the exercises. For now, don't worry about your grammar, spelling, and correct terms. In actual preparation of inspection reports, correct words are all generic terminologies, and you will be able to be familiar with those terminologies through frequent practice; in fact, references will be available for you. It could be your company safety policy, material handbook from your training, or from the regulatory agencies and standard organizations available online. In lesson 5, I will mention some of my techniques and use them to advise new safety officers in the field to do the same in order to improve their writing and reading skills.

The previous exercise is called basic safety inspection at home or workplace. *Congratulations!* Keep it up.

Now if you know someone who is currently working as a safety professional in your place, you can ask them to verify your report and seek for advices; don't be shy. If you don't know anyone, you can check sample reports online; just type 'home safety inspection report sample' and compare your own report, then if you have some mistakes, redo the report, and follow the example or instruction.

Lesson 5

IMPROVED WRITING AND READING SKILLS AT HOME

Tip 1: Read every day, as fast as you can. If safety-related reading materials are not available at home, purchase them at the bookshop or download them from the Internet and print them out. By using safety-related books as reading materials, you will be able to familiarize the safety terminology usually used by safety practitioners when preparing reports. If reading is not your habit, then make it your habit from now on. You might be experiencing difficulties, but it doesn't matter for now. For those experts in reading, all you need to learn and understand is that safety terminology.

Tip 2: Buy a blue notebook and start writing and transfer those materials you've read. Just like what we used to do in school, we used to write every day, copying what our teachers wrote on the blackboard. Am I right? The objective of this exercise is for you to easily remember and understand those words. If there is any drawing and table or sketch, copy those too. Within three months of doing this exercise, you will improve your reading and writing skills; just keep going, and put some reading materials in your pocket all the time.

Tip 3: Type what you have written in your blue book. If you are computer literate, then this exercise could be so easy for you, but still you need to do it. But for those who are not computer literate or not experts in using computers, then it's time for you to learn how to use the technology and be familiar with the basic Microsoft Word and Excel. Seek help from your friends and relatives to teach you. Continue this habit and exercise until such time that you're familiar with the safety terminologies and eventually improve your typing skill. This exercise is important because reading with understanding and preparing reports is one of the important everyday tasks of a safety practitioner.

Tip 4: Read some product catalogue (manuals) safety information available at home. You can find them inside the boxes of your appliances or equipment, or they are posted on the household chemical bottles or medicine boxes.

Tip 5: Purchase a book about developing reading and writing skills or watch YouTube videos; more techniques and detailed guidelines on how to improve your writing and reading skills will be in those tools. Tips 1, 2, 3 and 4 are tips based on my personal experience, but I can assure you that those are effective techniques.

Lesson 6

THE SAFETY PROFESSION IS MY LIFE NOW

As part of this book, I believe that giving advice and sharing knowledge and skills to others and to those new in this profession is the role of a good safety professional leader.

The title of this lesson, 'The Safety Profession Is My Life Now', is from the message of my colleague Sir Peter Q. Panganiban. He is currently working as safety supervisor with AECOM Arabia Ltd. Co., Saudi Arabia, and he has more than twelve years' experience as safety professional. He opted to work as a full-time safety officer in 2001 when his boss gave him an opportunity to handle a safety position. Do you think his boss would have given him a chance if he hadn't attended safety trainings prior to that opportunity and had the desire to be a safety professional someday? So he grabbed the opportunity and became one of the active members of PSSP-MER and eventually became an officer of PSSP-MER and improved his knowledge, skills, and capabilities about the fundamentals of safety. His primary reason was for career change and personal growth, and he says, 'I have no regrets.' Prior to his work as a safety officer, he was working as a technician and earning the ceiling salary grade of a technician, but now his monthly income has increased more than 300 per cent higher than his previous paycheck.

This was my reply when I received his response to my invitation to send me his message to be included in this book. I said, 'Wow! Very impressive principle, very touching, and a one-in-a-million message, bro, thank you so much. I assure you that you will be one of those individuals who will receive a copy of my manuscript once it's released. Once again, I am so thankful to have you as my friend, brother, and mentor.'

Below is his message for you to understand his principle and his characteristics as a safety professional. I hope you will be inspired and learn from it.

> Dear all,
>
> If one will ask me what it is like to be a safety practitioner or safety professional, well, for me, being a safety professional is not just a title or an ordinary job—it is an exceptional undertaking and a worthy profession. I am proud, and I really love being a safety professional. It is my life now. I save lives and maintain their well-being. I protect property and operations. I care for the environment and keep it sound. I make and share smiles with the personnel and their families when I implement and uphold the systems that safeguard and sustain the safety of the workers while working and letting them go back home happily and harmless. I create a safe, healthy, and grateful working environment. I contribute to building an exultant family and community.
>
> In all honesty, I must say also that being a safety professional is not an overnight title or profession. I have my principle founded on my own experience that is beneficial to share with everyone. It is not a secret and a perfect formula needs to be configured according to individual perspectives or attitudes. The principle states that being an effective safety professional actually requires credentials, competencies, exposures, initiatives, and positive attitudes. In other words, I am a safety professional not because I just know someone

but because I know something. I am a safety professional because I have qualifications, aptitudes, and experiences that are credited not just by paper but through my actions and professionalism. I am a safety practitioner because I do not want to be just a boss but to be a leader to guide. I am a successful safety professional because I constantly strive for competence and expertise in my chosen field. I do my best to be a mentor who influences others and adopts the lessons of safety in handling my own family, and I maintain being a worthy individual that knows how to respect and care.

Being a safety professional is my life now, and it will always be.

Peter Q. Panganiban
Safety Supervisor
AECOM Saudi Arabia Limited Company

Don't think that you will not be able to perform the duties and responsibilities of a safety practitioner. In every profession, there is a level of specific assigned responsibilities in each and every team member, just like in other fields or professions where the company allocates personnel in every team or unit to run the business. For now, just learn the basic duties and responsibilities of a safety officer like conducting daily safety inspections, identifying and resolving unsafe acts or unsafe conditions in the workplace, and you may participate in an incident investigation. As a safety officer you have to lead by example in complying with the company's safety policies, rules, and regulations; conduct toolbox meetings; and report near misses, accidents, and unsafe acts/conditions—this will be your important daily task as a safety officer.

Lesson 7

THE IMPORTANCE OF BEING A MEMBER OF A SAFETY ORGANIZATION OR SOCIETY

Twelve years ago, I really didn't know how and where to start. I strove to learn a lot by reading and asking questions to my colleagues and supervisors and all employees around me. Unfortunately, I was always not included in the company list every time they sent candidates for safety training in the Philippines. But I didn't stop or lose hope; instead, I kept performing all tasks assigned to me and continued praying to God and asking for help, and I kept my desire to learn in my own ways and always remembered my reasons, hoping someday that I would be working abroad as a safety practitioner. In July 2003, I was hired to work abroad as fire safety officer of Al-Mana General Hospital, Al-Hasa branch, Saudi Arabia. Thanks to God, God is the Greatest, here it is—my prayer has been answered. I am so thankful that I had very supportive supervisors in my first company in the Kingdom. They extended a lot of support and guidance for me to develop my knowledge and communication skills; they gave me a chance to conduct new employee safety orientations and conduct fire drills in the hospital and employee accommodations.

Thanks to God, I had about three years and a half working as a safety officer with my first company, earning an average monthly salary and was able to find a way to be a member of the PSSP-MER and had the opportunity to attend our regular training and seminars conducted every last Friday of the month. PSSP-MER is a non-profit organization and duly recognized member of the Accredited Community Partners (ACP) of Philippine Overseas Labor Office (POLO) and Overseas Workers Welfare Administration (OWWA) of Eastern Province, Saudi Arabia. Some of the goals and objectives of PSSP-MER are to promote, demonstrate, and lead best practices in the fields of safety, health, and environmental protection to promote continuing education and enhancement of knowledge and skills of the member in the fields of safety; health and environmental management trough trainings, seminars, conventions, consultations, and awareness campaigns; and other feasible means.

I was so lucky because one of my supervisors was so active in attending safety training, so I always had the opportunity to ride with him in his car. So within this period of time, I was able to obtain more than twenty credential certificates and develop and expand my knowledge, skills, qualifications, and self-confidence until such a time that I got a better opportunity to work with RSAL, which now is AECOM Ltd Co., and was assigned to join the project management team with the world's biggest oil company, the Saudi Aramco Company. I worked for about six and a half straight years from 2007 to mid 2013. God is the greatest. Thanks to Engr. Omar Rey Barraca for his recommendation seven years ago and to all my former bosses who help and guide me on my journey to be a successful safety professional like them.

Yes, I am promoting our society, the PSSP-MER, and all safety organizations around the globe that provide such services and trainings. This is your assignment to find the nearest safety organization, training center, or institution offering safety trainings. I encourage you to enroll yourself and join a safety organization in your city so you can start your journey to become a successful safety practitioner. If you are a Filipino citizen and are currently working in the Kingdom, you are welcome to join our society; Saudi nationals are also welcome to join the society. For more information on how to join our society, just Google *PSSP-MER* and contact any of the officers and BODs. As part of this lesson, please see below my original composition for our society. Hope you will find a good message in it. The PSSP-MER hymn is usually sung by the graduates of the Basic Occupational Safety and Health (BOSH) course during graduation day.

PSSP-MER Hymn
Original composition by Eisa Jesus G. Pedines Jr.
and Wife Jingle L. Pedines

I. I am proud member of PSSP-MER
Philippines Society of Safety Practitioners
Middle East Region, Saudi Arabia
My beloved society.
II. Uplifting safety wherever we are
Promoting health, safety & environment
Companies and communities
Needing our excellent services.

Chorus
PSSP-MER my society and second family
Founders thank you, I am so lucky
For having genuine friends, brothers
And committed officers and BODs.

Bridge
Our society a unique organization
Provides continued education
and supports to all its members,
a group of qualified safety practitioners
Our PSSP-MER,
Repeat Chorus

Lesson 8

PERSONAL KPI CHECKLIST

Here is my personal experience on the importance of writing and taking notes. Well, first of all, I am not a professional writer nor have I dreamed to be an author of a book—yes, I am an amateur songwriter maybe. How did I become the author of this book? This idea was developed after reading a book recommended by my bro Dennis Cullar. I will share more details on how I developed this handbook in the second edition, God willing. Anyhow you don't need to be perfect or a professional writer when writing down your personal checklist notes at home or at the workplace. This is my habit, taking notes; maybe you are already doing it but you are not aware of the importance of it. For me, as long as you understand what you wrote on a paper, whatever it is, it will help you to recall and guide you to complete a certain task that needs to be done first and last. Usually you can see those notes stuck on the walls or at home on the fridge. Take a look at my simple note below. The message below is a challenging phrase that I want to do, so I write on the Post-it and keep it in my pocket for me to remember that important message and do the necessary action when I settle my plan. The message below was a part of the speech of our PSSP-MER BOD chairman Engr. Peter M. Marfa, encouraging us to start developing our own Key Performance Indicator (KPI).

All companies have their own KPI targets; there might be different KPIs in every unit or department and organization with respect to the line of business they are in, wherein their success can be defined in terms of making progress towards their strategic goals.

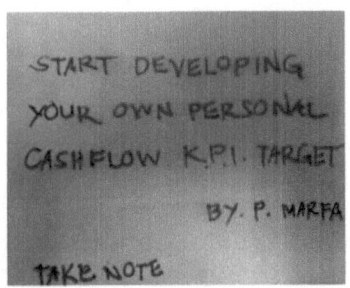

In my personal interpretation, we as individuals should have our own KPIs to set our targets and goals for ourselves or for our family such as children, education, monthly savings, investments, number of trainings, what and how many books you need to read, what age you are planning to retire, and when you want to have your own business, house, and lot and so on. To me, personal KPI is unlimited; you can list whatever target you want to achieve or want to have or to obtain. Setting targets starts from our thoughts, so *write it down*—don't just memorize it—and pray for it; remember lesson 1. Planning will follow as you regularly read your reasons and your personal KPI targets. Then when the plan is ready, follow your steps from A to Z as per your plan and pray for it regularly. Keep your burning desire, and do the required *positive* action according to your plan.

Yes! Writing down your personal KPIs and making a plan is very important but needs a positive approach and positive action. No matter how hard it is, start now and keep going; don't quit achieving your goals. Don't be afraid of failing; learn from your own mistakes and the mistakes of others as Napoleon Hill did. If you want to study the secrets of successful person on how they were able to conquer fear and challenges and become successful people, then grab the book of Napoleon Hill entitled *Think and Grow Rich*. This is such an amazing book as a guide, and it really works. I read this book several times, and I would love to read it again and again and again. Right now, I know I am on my way to having at least some of those secrets to being successful in what I am doing, and I will use and try those proven formulas as long as I am alive. In God's will.

I would like to thank my brother in APO, Dennis Cullar, who encouraged me to read the books of Napoleon Hill, Jim Rohn, and Robert Kiyosaki. Thank you, bro—if you didn't recommend those books probably I didn't able to have a wisdom to create my own manuscript even as simple like this, I owe you a copy of this book. May we always be LFS.

Lesson 9

HOW TO APPROACH SAFETY VIOLATORS IN THE WORKPLACE

Dealing with workers at the job site is not that difficult for those experienced safety practitioners. There are a lot of different approaches or techniques in dealing with employees in the job site. But personally, I always use the friendly professional approach as much as possible. I am a friendly guy anyway. Just deal with it professionally.

Be aware of those violators in the job site; some of them really don't know the requirements due to inadequate company support in providing safety awareness training and safety orientations. Some workers also intentionally violate the rules because of poor safety culture in an organization—the management doesn't show any concern, leadership, or accountability to their employees pertaining to a safe workplace and providing adequate safety supervision or providing proper personal protective equipment (PPE) and access to welfare facilities, for instance. They know that they themselves are committing violations or unsafe acts, but they don't care about it. 'Come what may,' they said. In my experience and knowledge gained from advanced courses, I learned that the usual reasons of those employees for continually violating rules are poor management; statements of intent not being in place; poor commitment to implementing HSE policy; no clear goals and objectives of the organization; inadequate allocation of safety responsibilities within the organization; understaffing of safety personnel, trainer, and supervisor, etc.; and unavailable procedures, processes, and guidelines on how to implement the HSE policy such as inspection procedure, work permit system, training requirements and procedures, and so on.

In my experience, I have encountered workers saying, 'Well, I have already worked many years in this company and nothing has happened to me. I am still alive.' Some say 'Using PPE is only a distraction when I am doing my job.' Others say, 'I don't care about it, our company doesn't care about our safety anyway.'

Well, those scenarios at the workplace are challenges to all safety practitioners in the job site. We cannot blame those individuals; most of the time, the best thing I've done was to be their friend, always smiling with them until such time I have a chance to explain to them the importance of obeying safety rules on-and-off-the-job site.

Other contributing factors are the following: they know someone in the management level as their backup, some employees just want to give you trouble because personally they don't like you, most of the time they are just following the instructions of their supervisors, and the last but not the least, poor awareness of the management about the importance of allocating adequate fund for safety implementation, which is one of the root causes of incidents and accidents in the workplace.

Now can you imagine if a new safety officer encounters such types of employees in the job site? I hope this book will help as a warning or build awareness of possible scenarios that you might encounter in the future. Those individuals representing the example scenarios above are employees that I have actually encountered in the workplace, not only once but many times. But this could be too much for a new safety officer to handle, couldn't it? You might cry in the corner of your office if you are sensitive and have weak determination. You might fight with them and your tears will drop, but my advice is don't quit. Be patient; just move on. Everything is gonna be all right; just pray for them.

Do you think can you handle such challenges? My answer is yes, of course you can do it. Trust me, because there are no special skills needed when dealing with employees committing violations in the job site, you can learn your own approach. It's just a matter of being passionate and being determined in what you are doing. Just maintain your burning desire; be familiar with the requirements that need to be done or are supposed to be done. Requirements are always a given; they are the in-place government rules/laws, regulatory agencies' requirements, standard organization requirements, client safety policies, or your company safety policy rules and regulations, so just be familiar with those requirements and lead by example that it's nothing more, nothing less. Nowadays, good safety culture is growing globally. The government, regulatory agencies, standard organizations, societies, institutions and companies, and private sectors are helping hand in hand to prevent, eliminate, or minimize tons of occupational diseases in the workplace. Companies are becoming more proactive in performing their moral, economic, and legal obligations and becoming extremely aware of the negative impacts on their businesses if safety, health, and environment issues were not properly managed.

So before you continue reading, I would love to ask you to write down below some of your personal approaches if you put yourself in a situation dealing with such violators in the job site. If you are currently working as a safety practitioner, this exercise will be easy for you. Just write down what actual actions or approaches you applied in handling such cases in the workplace.

My approach will be . . .

_____

_____

_____

_____

_____

Your personal approach written above could be better than my approach techniques, but if you think you need improvement, try to apply my personal techniques mentioned below.

Contributing factors and considerations that you need to feel are very important. You need to detect employees' facial reactions when you call their attention. For example, is he/she afraid, angry, or feeling sorry for what he/she has done? Is the employee feeling well? Is he/she a new employee in the workplace? Is he/she under pressure by his boss? Are they receiving their salary on time? It might be three months delayed. Employee might have lack of sleep or might be facing family problems.

Now here are my tips and personal advice when dealing with such employees:

1. Rule of thumb—show respect so you will be respected in return, and extend your greeting to them. Offering a handshake will be better.
2. Smile and introduce yourself professionally. Then if you know or feel any unusual situation listed under 'Contributing Factors and Considerations' that may affect their normal behavior when talking to you, be careful to not offend their feelings and try to address them. If you have a chance, let them feel that you care about their safety and health and their feelings.
3. Then discuss professionally your concern. If it is an individual violation, then talk to the involved employee alone, but if it is a group violation, then gather them together and conduct a toolbox talk as one and provide them some awareness about the requirements and the violation they committed. If possible, let them understand the consequences that may occur if they continue doing such unsafe acts or unsafe conditions.
4. Be humble. Understand and listen to their concern, explanation, or justification, but it doesn't mean that you should compromise their safety.

5. If they argue, do not make any conversation or fight with them. Rather, discuss with your supervisor what happened, and the best thing you can do that moment is to pray for them. Always remember the power of prayers to our Almighty Creator.

Repeated violations must be reported and corrected as soon as possible to avoid any further near misses or accidents and property damage in your area of responsibility.

The time will come, sooner or later, when you will improve and develop your communication skills and be knowledgeable on how to handle such cases or situations in the workplace. Just keep your desire to learn more and continue practicing problem-solving skills in addressing those deficiencies you've encountered in the job site, and be friendly and be patient. Learn to respect others so they will respect you in return. Be active in your safety society or organization, and seek more advice from your colleagues and mentors.

Lesson 10

SAMPLE PHOTOS OF BASIC UNSAFE ACTS, UNSAFE CONDITIONS

In this lesson, you will be able to be familiar with some actual findings in the construction site. Each item has findings and a photo. On the top of each item, basic information was provided for the sake of learning company names (A, B, C, or D), person concerned (Mr X, Y, or Z), date, and exact location (Location 1, 2, 3, and so on). Details below the picture are samples of how to write a finding, recommended corrective action or control to be taken, and the potential consequences or risks if the hazard condition has not been corrected in a timely manner. Remember in your recommendation to always add a reference based on your future company's safety policy or any standard requirements from a regulatory agency such as OSHA or ILO or local government safety requirements or as per any standard organization such as ANSI, NFPA, UL, CE, etc.

I decided not to put any specific standard requirements or references on sample reports to avoid any conflict because companies set their own safety standards and requirements within their organization in line with the government safety and health requirements or international regulatory agency requirements.

Also, the reason I didn't put the corrective action photo is basically for your learning purposes. Do your best to analyze how to resolve the unsafe condition based on my recommendation.

Completion date of the corrective action usually added in the end of the recommendations it depends the risk level and category. Remember that your recommendations should be specific, measurable, achievable, realistic and time-bound. Completion date can be; as soon as possible or immediate; within a day or days; week or weeks; a month or months; or put the exact date. For now, try to figure out how and what is supposed to be done and what it looks like when corrective action has been put in place and decide your own time-bound target date just for learning.

Sample Observation 1

Company name: B	Exact location: Loc. 1
Person concerned: Mr X	Date: MM/DD/YEAR
Activity: N/A	Corrective action photo

Observation/findings: Unsafe condition; improper platform installation, plywood and unapproved planks were used. This is a result of inadequate competent scaffold supervisor during erection.

Potential consequences/risks: Workers may fall from heights, scaffold tower collapse, falling objects.

Recommended corrective action/control: No Entry warning shall be posted on the ladder access of the platform. Rectify the scaffold platform; remove unapproved planks, and use approved materials as per company requirements. Rectification of the scaffold shall be supervised by a competent person and must be reinspected prior to use.

Sample Observation 2

Company name: B	Exact location: Loc. 2
Person concerned: Mr Y	Date: MM/DD/YEAR
Crane and rigging activity.	Corrective action photo

Observation/findings: Unstable crane positioning

Potential consequences/risks: Crane may tip-over and may lead to property damage and personnel injuries.

Recommended corrective action/control: Soil shall be compacted properly or provide proper wooded planks pad for the outrigger pad; pre-operational crane inspection shall be carried out prior to any lift. Crane operator retraining is recommended and ensures adequate safety instruction and supervision.

Sample Observation 3

Company name: B	Exact location: Loc. 3
Person concerned: Mr Z	Date: MM/DD/YEAR
Mechanical concrete chipping—What are missing PPEs for this activity?	Corrective action photo

Observation/findings: Unsafe act; workers not wearing eye protection and hearing protection.

Potential consequences/risks: flying objects or ejection of concrete particles may cause eye injury and exposure to excessive noise may induce hearing lose.

Recommended corrective action/control: Stop work and provide proper PPE (e.g. eye goggles and face shield and ear plug; retraining on PPE requirements shall be carried-out immediately.

Sample Observation 3

Company name: B	Exact location: Loc. 4
Person concerned: Mr Z	Date: MM/DD/YEAR
Activity: Civil works	Corrective action photo
Observation/findings: Unsafe condition; poor cable management, poor housekeeping. **Potential consequences/risks**: Personal injury, trip/fall and electrocution hazard. **Recommended corrective action/control:** Improve cable management; electrical cable shall be elevated to prevent potential cable damage, electrical hazard, and tripping hazard.	

Sample Observation 4

Company name: B	Exact location: Loc. 5
Person concerned: Mr H	Date: MM/DD/YEAR
Working in the dark.	Corrective action photo
Observation/findings: Unsafe condition; inadequate lighting in the workplace. **Potential consequences/risks**: Slip/trip/fall, impalement, being struck on, or struck by unseen protruding objects and may lead to personal injury. **Recommended corrective action/control:** Provide sufficient illumination in this area and other areas in the building.	

Sample Observation 5

Company name: C	Exact location: Loc. 5
Person concerned: Mr E	Date: MM/DD/YEAR
Activity: Carpentry works	**Corrective action photo**

Observation/findings: Unsafe act/condition; worker using damaged hand tool, inadequate hand tools inspection, failure to follow procedure.

Potential consequences/risks: Personal injury.

Recommended corrective action/control: Remove the damage hummer from the job site imedaitely; reinspect all hand tools in the warehouse prior to issuance.

Sample Observation 6

Company name: E	Exact location: Loc. 6
Person concerned: Mr E	Date: MM/DD/YEAR
Activity: Mechanical works ducting installation.	Corrective action photo

Observation/findings: Unsafe act/condition; workers are not protected from the hazardous fiberglass insulation; inadequate awareness, and inadequate PPE supply.

Potential consequences/risks: Personnel are prone to respiratory diseases if not corrected.

Recommended corrective action/control: Conduct HAZCOM and respiratory awareness training, provide proper respiratory protection equipment, and ensure adequate safety monitoring, instruction, and work supervision, and conduct job safety analysis for this activity ASAP.

Sample Observation 7

Company name: C	Exact location: Loc. 7
Person concerned: Mr E	Date: MM/DD/YEAR
Access over a trench	Corrective action photo

Observation/findings: Unsafe access; plywood has been placed as temporary crossover access on the excavation.

Potential consequences/risks: Fall to lower level may cause personal injury.

Recommended corrective action/control: Remove the unsafe access and provide safe access or reroute access where crossover access is not required and provide adequate directional signage immediately.

Sample Observation 8

Company name: C	Exact location: Loc. 8
Person concerned: Mr E	Date: MM/DD/YEAR
Activity: Civil works	Corrective action photo

Observation/findings: Unsafe act/condition; working on the uninspected scaffolding and without using fall-arrest equipment.

Potential consequences/risks: Fall to lower level may cause serious personal injury.

Recommended corrective action/control: Stop work imediately; inspection of the scaffold platform is required prior to resuming activity; provide training to all personnel involved and provide fall arrest equipment to all workers working at heights; ensure proper supervision and safety monitoring. Training shall be completed on on before MM/DD/2013.

Sample Observation 9

Company name: C	Exact location: Loc. 9
Person concerned: Mr E	Date: MM/DD/YEAR
Is this a hole on the scaffolding platform?	Corrective action photo

Observation/findings: Damaged scaffold platform/planks.

Potential consequences/risks: Trip and fall hazard.

Recommended corrective action/control: Replace the damaged planks and reinspect the platform immediately.

Sample Observation 10

Company name: F	Exact location: Loc. 10
Person concerned: Mr AZ	Date: MM/DD/YEAR
Activity: N/A	Corrective action photo

Observation/findings: Unsafe condition; poor housekeeping, inadequate awareness on hazardous fiberglass insulation hazard.

Potential consequences/risks: Potential inhalation of hazardous dust may lead to respiratory disease.

Recommended corrective action/control: Remove these materials as soon as possible, conduct respiratory awareness training, provide proper respiratory and body protection to those garbage collectors/housekeepers.

Sample Observation 11

Company name: F	Exact location: Loc. 10
Person concerned: Mr AZ	Date: MM/DD/YEAR
Activity: N/A	Corrective action photo

Observation/findings: Unsafe condition; unsafe access, protruding rebar along the access.

Potential consequences/risks: Slip/trip/fall, impalement hazard, or struck on an unprotected rebar along the unsafe access.

Recommended corrective action/control: Close this temporary access or provide proper access with hand rail; if rebar is needed for civil marking purposes then caps on the protruding rebar must be provided ASAP

Sample Observation 12

Company name: F	Exact location: Loc. 12
Person concerned: Mr AB	Date: MM/DD/YEAR
Activity: Working in confined space/drilling.	

Observation/findings: Unsafe act/condition; working in a confined space without Confined Space Entry (CSE) permit. No ventilation, no proper access, no log in/out procedure.

Potential consequences/risks: Immediately dangerous to life and health (IDLH) situation.

Recommended corrective action/control: Stop works immediately; provide training to all personnel involved in confined space entry activity, and ensure proper supervision and work permit system procedure shall be in place, including gas testing prior to entry in any confined space. Training shall be conducted within this week.

Lesson 11

RECOMMENDED SAFETY TRAINING FOR NEW SAFETY OFFICERS

Hopefully you've already tried checking online some safety organizations or safety training centers in your places. Well, I am impressed; you've reached this far reading this book. If you think it is beneficial for you and others, you may recommend it to your friends and relatives.

In this lesson, considering your knowledge and understanding capabilities on safety-related topics, recommended safety subjects suitable for a new safety officer will be mentioned, and it is better for you to take the following subjects. First of all, the best course for you is a safety introductory subject like the Basic Occupational Safety and Health (BOSH) course in construction. Upon completion of this course, you will be able to understand the fundamentals of safety in different topics. Several training institutions/organizations or government agencies are offering this course, such as Occupational Safety and Health Center (OSHC), Philippines, and the Safety Organization of the Philippines Inc. (SOPI). SOPI was my first safety organization in the Philippines, and I am still active in renewing my membership every time I visit their main office. If you are in Saudi Arabia, please contact PSSP-MER if you want to join our society. For other nationalities, check it online; a variety of training institutions provides safety training.

I took my NEBOSH IGC on 2013 at Riyadh, Saudi Arabia, with a local accredited training provider. I decided to take the NEBOSH IGC to upgrade my credentials, so I used Google search engine and found a training provider that offered night class, and I did it; I passed NEBOSH IGC with credit. NEBOSH courses can be taken around the globe. I also have several OSHA trainings sponsored by Saudi Aramco and provided by the US Department of Labor OSHA Training Institute, Region IX Education Center, a division of UCSD Extended Studies and Public Programming, so check UCSD online if you wish to take OSHA courses. For fall protection–related courses, GRAVITEC is a well-known training provider. I've been certified as a competent FP trainer and a FP-competent person by GRAVITEC through the Saudi Aramco training program. In your situation, you have to invest some funds or cash from your own pocket if you are currently working with a company not sponsoring safety staff on such higher-level safety courses to upgrade your credentials.

The other way around, taking fundamental safety by subject basis, is also recommended such as electrical safety, fire safety, basic safety inspection, hazard identification, working at heights safety, job safety analysis, safety inspection, crane and rigging safety, housekeeping safety, basic incident investigation and reporting, confined space entry procedure, work permit system, excavation safety, manual handling safety, ergonomic safety, driving safety, road safety, behavioral safety, etc.

Those recommended specific subjects are necessary for you to understand basic safety fundamentals prior to undergoing any advance safety training certification or accreditation such as NEBOSH courses, OSHA courses, and a variety of advanced train-the-trainer courses provided by other well-known training providers (government sectors or agencies, organizations, institutions, or societies). Once again, check it online and find the nearest safety institution and organization in your places.

In my opinion, taking any advanced safety courses or certification could be easy to pass if you have the burning desire to pass that chosen higher-level course. For those individuals who are currently practicing the safety profession, that is only an extra plus, but it doesn't guarantee that you will pass those higher-level courses if you don't study your lessons and are overconfident. That's why it is very important to study your lessons; after the classroom training, have hands-on practice, attend fundamental safety training, and read a lot of information prior to taking any advance courses to ensure that you have a chance to pass.

I recommend that after completing any courses, you need to go over your handbooks provided by your training provider again and again or apply my tips in lesson 5. Especially if you are currently working in a field other than the safety profession, every time you accomplish or obtain training certificates and credentials, you shall then update your CV and be familiar with all information you provided on your CVs. Now if you are currently jobless, it's better for you to find a job, whether a safety position or not, because at this stage, you need to finance your training or save some cash for the next course you intend to take. For those currently working as safety practitioners, make it sure that at least you have one or two training sessions attended each year to upgrade your credentials and CVs. For more references, use the Google Play store online; use the technology to help you expand your knowledge on the application of what you have learned from your trainings or study in advance information for your next training. Usually, most individuals download games, movies, and music, but for you, you need to use your spare time reading and watching safety videos and other safety-related information or topics available online. Just type the word *safety* in the search engine of the Google Play store, and you can download a variety of *free* useful information, applications, tips, videos, and other services about safety. So invest your time using the technology, this time to expand and speed up, gaining safety knowledge. So reduce some of your usual daily habits.

Lesson 12

MESSAGES FROM MY MENTORS

I have met and seen some safety practitioners who haven't attempted or have not able to climb the corporate ladder or have not been promoted to higher positions because of lack of confidence, knowledge, skills, and credentials, because they stopped learning and failed to take any advanced safety courses, certification, or accreditations. Read the messages from my three top mentors for you to be inspired like me; they are very successful safety professionals and pillars of our society.

Engr. Rey Omar Barraca is one of the founders of the PSSP-MER, currently the secretary general of the society, has thirty-one years of experience in a safety profession, and is currently working as HSE manager with AECOM Saudi Arabia Limited Company.

It was in 1983 when he was given the opportunity to work as training officer for safety, security, and firefighting crews at King Abdulaziz Military Academy in Riyadh, Kingdom of Saudi Arabia. He continued to work as a training officer until he became a safety, health, and environment leader. He constantly looks for new ways to improve and develop an environmentally friendly, safe, and sustainable workplace, and in a short period of time, he has made immense progress towards his goal, and it was successful. He said, 'Now I have thirty-one years' experience as a safety professional and still keeping safety at the forefront is a challenge for which success is crucial, making progress on achieving my goal, and I am looking forward to providing a collective knowledge with our PSSP-MER colleagues. I believe the best HSE practices created by our expanded capabilities, research, and continuous learning through training will achieve a safer and healthier working environment.'

Message 1

Safety professionals make decisions drastically during critical moments at the project sites simply because we want to maintain healthy and vibrant workplaces; protect facilities, equipment, and the environment; and most importantly, ensure the safety of the workers. We ensure that the workers are able to serve the company, clients, and even the community in a healthy and safe environment. For those who are practicing and wish to practice safety, you need more patience and dedication, hence our commitment to safety is constantly reinforced through innovative processes, procedures, and standards including comprehensive training as provided by PSSP-MER. Human activities reflect the universally accepted principles, standards, and even culture, beliefs, and traditions that focus on fairness, human health, environmental quality, ethics, and the advancement of society. PSSP-MER makes it clear that we genuinely care about the well-being of the members, and we are looking to continuously improve our current safety program with new and innovative technology as well as engineering processes and procedures. We utilize all the methods available to keep our family and even the workers of the company and the community always safe both on and off the job. Remember, dedication and sincerity in your profession is the great success in HSE.

Engr. Rey Omar Barraca
HSE Manager
AECOM Saudi Arabia Limited Company

The second message is from my former safety engineer / supervisor at Al Mana General Hospital eight years ago. He is currently working with the Saudi Aramco project management team as division safety coordinator and is currently the vice president of internal affairs of the PSSP-MER. He has unique and very strong principles in life and is an excellent leader of his team, a leader that produce future leaders, a very supportive and proactive supervisor. He is an electrical engineer by profession and has more than thirty years' experience in the Kingdom and has become a successful safety professional.

Message 2

Being willing to be a safety practitioner is not enough to succeed in this field of expertise. The mind and the heart have to work hand in hand. The heart is for accepting the responsibility that a safety practitioner has for the safety, welfare, and well-being of the worker in the workplace. One special virtue of a safety practitioner is the acceptance of the above moral obligation, which I believe is not so present in other fields or disciplines. The mind has to be sharp, accurate, and knowledgeable in the safety requirements of the worksite or organization. Know-how, training, and field experience can make or break the future of a safety practitioner. A safety practitioner should be hands-on in the areas of safety design, implementation, control, and in the evaluation for further improvements. The safety profession is not a tabletop profession; field and theoretical expertise are both required. Excellence in this profession cannot be achieved from either the office or the field. It needs adequate exposure in both venues.

My advice to fellow and future safety practitioners: Do not stop learning. The Internet is readily available as a source of knowledge in the safety profession. Be a member of a safety organization. It is your membership in this organization where you can compare your level of expertise and thus improve if found lacking by comparison to your fellow safety practitioners. Invest in your profession. Spend time, money, and effort, to uplift your knowledge and scope of expertise in the field of safety. Go to the field, practice what you learn, and observe the practice of safety by the general workforce. Aim to improve the safety culture as the final goal of your existence in that organization. Without a clear and sincere objective or goal in your career as a safety practitioner, failure is a common result. Financial stability as a main goal will fail. Financial stability is just a secondary result and, in the field of safety, is always a good result but should never be a goal.

As a safety practitioner, I shall do my best to ensure all the workers within my responsibility will go out of this workplace safe and alive, ready to come to work the next day. A very simple objective but it carries a lot of obligation and responsibility *only a true safety practitioner aims to fulfill.*

Engr. Levi S. Alejo
Division Safety Coordinator
Saudi Aramco Project Management Team

The last message is from our PSSP-MER BOD chairman of the board, who has twenty-one years' experience as a successful safety professional, an individual that has wonderful wisdom and an advocacy for helping people by being an active Filipino community leader in the Kingdom and in the Philippines. He has a masteral degree and a doctoral degree in business administration. I hope you will get the very important tips from his short message.

Message 3

For an individual who wishes to shift from the present job to the safety profession, it is a big challenge for their career. However, he has a big opportunity to find a better job and compensation in the future competition since there is no safety engineering course in our country and the safety profession is highly in demand in the market at present. The only chance to become a safety practitioner in the near future is to grab the opportunity to attend safety courses given by any international safety organization and learn the safety standard requirements related to the present job. As soon as an individual obtains enough knowledge about the roles and regulations in safety, then he may try to find an opportunity to apply the knowledge learned during the seminar and training presentations.

Engr. Peter Marfa
Safety Superintendent
Saudi Aramco Project Management Team

Lesson 13

A REMINDER WHEN YOU GET A JOB

If you already have the guts to apply, prepare your résumé and distribute to at least ten to twenty certified recruitment agencies in your area or any local company. Attending interviews will be another challenge for you to survive. It doesn't matter if you will be rejected or fail several interviews. Use the opportunity to learn from the questions of the interviewers. Remember all the questions asked during the interview; you need to write down all questions you remember and try to answer them properly in English at home in front of the mirror. Doing this exercise will eventually help you to build your confidence for the next interview. Remember that no matter how many rejections you will experience, just do it, and do not forget lesson 1.

Now once you get your job, remember to prioritize paying all your accountabilities to your family or friends as you promised them. Make a habit to extend your help to the needy; give charity in your own way. In Islam, even a smile is charity. There will be a time that you will feel so tired and unhappy and will almost want to quit or resign without proper planning. But once again, do not give up. Read lesson 1 again. Always remember your reasons and just keep your burning desire to overcome all challenges on your way. You might experience temporary defeat or failure, rejection, and criticism from other people or from your own relatives, but it doesn't mean 'game over'; you have just started your journey. Review your performance and persistence on performing your plan; read your reasons regularly and always apply lesson 1. It's okay to experience failures because failing is the key to success, so learn the lessons from those failures you've experienced. But if you really decide to move on and plan to find a better opportunity and a better working environment, then pray for it first prior to doing so. Prepare a plan; do the same steps, especially lesson 1. As Robert Kiyosaki says, 'If you've failed, that means you're doing something, and if you are doing something, you have a chance.' Robert Kiyosaki is a Japanese American successful investor and author of the popular book *Rich Dad Poor Dad*, where he wrote about his two dads. I would love to encourage you to purchase that book and his book entitled *Before You Quit Your Job*. Those books will be your guide to level up your mindset a few years from now once you've stabilized your financial status from being a safety practitioner. Hopefully once you are receiving a good salary grade, I hope that the next level you should think about is to start your own business two to five years from now. There is a lot of advice from Mr Kiyosaki that I've tried. In my personal experience, after reading some of his books, it pushed me to keep working hard and read more books to level up my mindset. We started four traditional businesses in 2013 in the Philippines. We experienced and learned how to fail and to be broke but became smarter in spending money. This is all part of the process that we need to experience—to become temporarily defeated and look forward to being better entrepreneurs in the future. So my point of view here is that it's okay to be an employee for a certain period of time. Also, if you are happy to retire as an employee, go for it, but if someday you want to start your own business, you can start learning now by reading books of those successful entrepreneurs and investors to inspire you and upgrade your mindset. For us, me and my loving queen, my wife, Hessa Jingle, we are still in the process of mastering Robert Kiyosaki's formula on how to become a successful entrepreneur. Robert Kiyosaki's formula is amazing; it is just a matter of changing your mindset. For me, working as an employee with a settled mindset is okay as long as

I am happy until such time that I want to try my wings on a different level. If you are in the Philippines, then you can attend similar training conducted by Filipino motivational speakers like Chinkee Tan or Bo Sanchez. They are Filipino motivational speakers and best-selling authors in the Philippines. In 2013, I attended Chinkee Tan's seminar conducted in Baguio City. If you are in the Philippines, do your best to find ways to attend their seminars, and start reading successful stories. Open your mind, and start learning the secrets of successful individuals, and be inspired.

In our current generation, the twenty-first century, you cannot be employed as a teacher, nurse, and accountant and so on if you don't have or possess a bachelor's degree diploma. In my opinion, aside from being an entrepreneur, the safety profession also doesn't require a bachelor's degree. Although it is an advantage and an extra plus to have a degree diploma, the most important thing in this field is to have dedication to your job and to have the desire to learn. Invest time and money to attend training to obtain credentials. Be involved and be updated on current safety trends around the globe by being a member of safety organizations to be able to share and expand your knowledge, skills, and abilities. Positive desire combined with a detailed plan and positive action will definitely give you positive result.

I hope to see you someday in your places, God willing, or see you in our future online game (Ladder Safety Journey Game), which will be available on my website. This is one of my visions: to make an online reality game for safety practitioners to share knowledge and help new safe practitioners to enhance and practice abilities, skills, decision-making, reporting, hazard identification, leadership skills, and so on. See the future game screen shot.

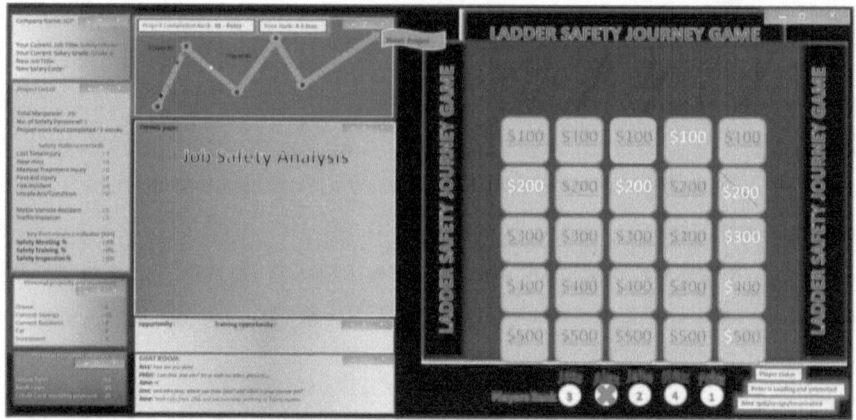

I provided additional exercises as the last part of this handbook; please try your best to complete the exercises. You can redo the report as many times as you want using a blank piece of paper or use a pencil if you decide to write on the book until such time that you develop your analysis and reporting skills.

Please be reminded that all the sample reports are related to construction job site unsafe acts and unsafe conditions, but it doesn't mean that your duties and responsibilities are only in the construction sites. Being a safety practitioner, you will be tasked on the other areas of responsibilities such as monitoring driving and road and traffic safety at the job site. You will be inspecting the basic welfare facilities such as water stations, proper housekeeping and hygiene, and maintenance of toilets, dining areas, rest areas, smoking shelters, employee accommodations, offices, and clinics. Regular inspection of power tools and safety equipment will be part of your scope and responsibilities. You might be given a task by your supervisor to conduct emergency drills, and as you level up your position in the future, you will handling projects and will be assigned to lead a team and be involved in developing company safety policies, conducting risk assessments and job safety analysis, and providing training. You might be involved in project design and the review of a project's scope and safety requirements. Record-keeping will be your responsibility such as maintaining the leading and logging indicators such as KPI records, safety statistics, incident reports, minutes of meetings, inspection reports, and so on.

Peace be with you, and may our Almighty Creator guide you and help you to fulfill your goals in life.

Be safe, and love and take care of your family.

Lesson 14

UNSAFE ACT AND UNSAFE CONDITION REPORT EXERCISES

Remember that using your own words when preparing your report is a good exercise to improve your writing skills. As much as possible, avoid the copy-and-paste technique unless it is required to copy the exact phrases from a certain references. For now, it is better for you to understand the requirements and use your own sentences or paragraphs.

I prepared some more incomplete observation reports for you to complete. This exercise will be a preparation for your exercise in the last part of this handbook. For those safety practitioners, you can add references from your own company safety policy or from any enforcement agency and standard organization requirements. Missing details on the report are in bold font. You can create your own area, project name, and person concerned and put any date you want and in end of your recommendation put completion date.

Exercise 1

Company name: F	Exact location: Loc. 10
Person concerned: Mr AZ	Date:
	Corrective action photo

Observation/findings: Unsafe condition/unsafe act; taking short-cut workers passing through the unsafe access.

Potential consequences/risks: Workers may fall into the trench that mey leads to personnel fatal accident.

Recommended corrective action/control:

Exercise 2

Company name: F	Exact location: Loc. 10
Person concerned: Mr AZ	Date:
What is the problem with here? [photo]	Corrective action photo

Observation/findings:
_____
_____

Potential consequences/risks: Personnel injury due to Soil collapse or soil erotion; workers may fall into the trench.

Recommended corrective action/control: Secure the excavation by provid barricade or exclusion as soon as possible; Adequate warning sign shall be provided immediately; Engineered shoring and proper acces shall be put in place prior and obtain confined space entry permit prior to resume activity on or before Dec.25, 2013

Exercise 3

Company name: F	Exact location: Loc. 10
Person concerned: Mr AZ	Date:
What is the problem with the access?	Corrective action photo

Observation/findings:
_____
_____

Potential consequences/risks: Tripping or falling to lower level could result to personal injury.

Recommended corrective action/control: Proper or safe access shall be provided immediately

Exercise 4

Company name: F	Exact location:
Person concerned: Mr AZ	Date:
What is the missing PPE on handling hazardous fiber?	Corrective action photo

Observation/findings: Unsafe act; not using proper respiratory PPE

Potential consequences/risks:
_____
_____
_____
_____

Recommended corrective action/control: Use approved PPE for handling fiberglass such as wearing coveralls and particulate respiratory protection; fiberglass hazard awareness shall be conducted and provide adequate PPE supply as soon as possible.

Exercise 5

Company name: F	Exact location: Loc. 10
Person concerned:	Date:
What is unsafe in this stair?	Corrective action photo

Observation/findings: Inadequate warning sign; stair under construction in unsafe condition.

Potential consequences/risks:
_____
_____
_____
_____

Recommended corrective action/control: Close the area with a hard barricade and provide warning signboards on both ends of the stairs ASAP.

Exercise 6

Company name: F	Exact location:
Person concerned: Mr AZ	Date:
Diesel fuel container—what is missing?	Corrective action photo

Observation/findings: Combustible liquid container not properly labeled.

Potential consequences/risks: Potential of the content being misused and may be stored with incompatible chemical or may be in contact with heat source and lead to explosion or fire incident.

Recommended corrective action/control:
_____
_____
_____
_____

Exercise 7

Company name: F	Exact location: Loc. 10
Person concerned: Mr AZ	Date:
Is a 'come-a-long' hoist safe to anchor a fall	Corrective action photo

Observation/findings: Unsafe condition/unsafe act; improper anchor point, 'come-a-long' devise is not fall protection equipment.

Potential consequences/risks:
_____
_____
_____

Recommended corrective action/control: 'Come-a long' shall not be used as part of a personal fall arrest system or anchor point.

Exercise 8

Company name: F	Exact location: Loc. 10
Person concerned: Mr AZ	Date:
Closely look at the picture. What is the hazard?	Corrective action photo

Observation/findings:

Potential consequences/risks: Puncture injury

Recommended corrective action/control: Used lumber shall have all nails removed before stacking or before disposal, or secure in the de-nailing area. Verify all similar situations all over workplace and provide safety instructions to the crew immediately.

Exercise 9

Company name:	Exact location: Loc. 10
Person concerned:	Date:
This is obvious—what is the problem?	Corrective action photo

Observation/findings: Poor housekeeping

Potential consequences/risks:
_____
_____

Recommended corrective action/control: Clean the area, and eliminate all waste and combustible materials to prevent fire incidents. Conduct daily toolbox talk and discussed the importance of good housekeeping; ASAP

Exercise 10

Company name: F	Exact location: Loc. 10
Person concerned:	Date:
Be specific on this picture. What is the unsafe condition?	Corrective action photo

Observation/findings: Poor housekeeping, storing combustible material under electrical panel board.

Potential consequences/risks: Fire hazard

Recommended corrective action/control:
_____
_____
_____
_____

Exercise 11

Company name:	Exact location:
Person concerned: Mr AZ	Date:
An open manhole? Yes, it is. What is the hazard?	Corrective action photo

Observation/findings:
_____
_____

Potential consequences/risks: Someone could fall in the confined space, which could result in major personal injury.

Recommended corrective action/control: Eliminate the hazard by installing the cover or secure it by providing a barrier, a hard barricade, and adequate warning sign. ASAP

Exercise 12

Company name: F	Exact location: Loc. 10
Person concerned:	Date:
An extension cord—what is dangerous in this condition?	Corrective action photo

Observation/findings:

Potential consequences/risks: Potential electrocution

Recommended corrective action/control: Remove the damaged cable, and get it fixed by certified electrician; inspection of all electrical cables and cords shall be conducted on a regular basis, and provide instructions to all users to conduct daily pre-inspection of cable prior to use. Supervisor should not allow damaged cable in the job site. ASAP

Exercise 13

Company name: F	Exact location: Loc. 10
Person concerned:	Date:
Electrical crew working in the dark?	Corrective action photo

Observation/findings:
_____
_____

Potential consequences/risks:
_____
_____

Recommended corrective action/control: Stop work and provide adequate lighting in the workplace; counselling with the supervisor shall be carried out to warn him not to allow his crew working in the dark.

Exercise 14

Company name:	Exact location: Loc. 10
Person concerned:	Date:
Is the crane stable?	Corrective action photo

Observation/findings: Unstable outrigger pad of the crane/improper setting of the crane.

Potential consequences/risks: Incompact soil will erode/collapse and may cause the crane to become unstable and tip over the crane that could lead to massive property damage and multiple personal injury and fatality.

Recommended corrective action/control:
_____
_____
_____
_____

Exercise 15

Company name: F	Exact location:
Person concerned:	Date:
	Corrective action photo

Observation/findings: Unsafe working flatform; fall hazard.

Potential consequences/risks:
_____
_____

Recommended corrective action/control:
_____
_____
_____
_____

Note: Answer can be in different approaches, but at least you have to use your own sentence. There is a variety of best-practice control measure recommendations from enforcement agencies or from standard organizations.

Below are my possible answers for exercises 1–15:
Do your best to answer exercises 16–71.

Answer for exercise 1.

Potential consequences/risks: Workers may fall into the trench that mey leads to personnel fatal accident.

Recommended corrective action/control: Close this area, provide No Entry warning sign immediately; Condcut emergency meeting to all supervisor and discuss the unsafe condition tomorrow MM/DD/2013.

Answer for exercise 2.

Observation/findings: Excavation cave-in and inadequate warning sign board and barricade on the edge of the excavation

Answer for exercise 3.

Observation/findings: Unsafe condition, improper access.

Answer for exercise 4.

Potential consequences/risks: Exposure to hazardous fiberglass dust from cutting ducting insulation may leads to respiratory illnesses or lung diseases.

Answer for exercise 5.

Potential consequences/risks: Workers may use the unsafe stair; personnel may fall to lower level and cause injury.

Answer for exercise 6.

Recommended corrective action/control: All chemical containers shall be properly labeled and protected and kept in a safe designated area.

Answer for exercise 7.

Potential consequences/risks: using improper fall arrest equipment or anchor point may result to personal injury.

Answer for exercise 8.

Observation/findings: Projecting nails along the walkways.

Answer for exercise 9.

Potential consequences/risks: Fire hazard.

Answer for exercise 10.

Recommended corrective action/control: Clean the area and eliminate all waste and combustible material to eliminate fire hazard. Electrical panel board/equipment shall be free from any obstruction and combustible materials.

Answer for exercise 11.

Observation/findings: Unsecured manhole, inadequate warning sign and barricade.

Answer for exercise 12.

Observation/findings: damaged extension cable being used; inadequate electrical cable inspection prior to use.

Answer for exercise 13:

Observation/findings: Unsafe condition, very poor illumination in the workplace.

Potential consequences/risks: Potential struck-by or caught in between, slip/trip/fall hazard due to poor illumination in the working area that may lead to personnel injury.

Answer for exercise 14:

Recommended corrective action/control: Stop work and conduct a meeting with the lifting crew and conduct an investigation why they would allow this very unsafe condition on lifting activity; Soil shall be compacted properly or provide wooded planks pad for the outrigger pad, pre-operational crane inspection shall be carried out prior to any lift.

Answer for exercise 15:

Potential consequences/risks: Substandard or improper working platform may collapse and may cause serious personnel injuries.

Recommended corrective action/control: Dismantle the unsafe working platform and used approved scaffolding.

Exercise 16

Company name:	Exact location:
Person concerned:	Date:
Is it safe to sit on the top of an A-type ladder?	Corrective action photo

Observation/findings:
_____
_____

Potential consequences/risks:
_____
_____

Recommended corrective action/Date:
_____
_____
_____
_____

Exercise 17

Company name:	Exact location:
Person concerned:	Date:
Open manhole/shaft?	Corrective action photo

Observation/findings:
_____
_____

Potential consequences/risks:
_____
_____

Recommended corrective action/Date:
_____
_____
_____
_____

Exercise 18

Company name:	Exact location:
Person concerned:	Date:
What is missing in this excavation?	Corrective action photo

Observation/findings:
_____
_____

Potential consequences/risks:
_____
_____

Recommended corrective action/Date:
_____
_____
_____
_____

Exercise 19

Company name:	Exact location:
Person concerned:	Date:
Site access road shoulder—is it safe or unsafe?	Corrective action photo

Observation/findings:
___
___

Potential consequences/risks:
___
___

Recommended corrective action/Date:
___
___
___
___

Exercise 20

Company name:	Exact location:
Person concerned:	Date:
Fabricated ladder—is it acceptable?	Corrective action photo

Observation/findings:
_____
_____

Potential consequences/risk:
_____
_____

Recommended corrective action/Date:
_____
_____
_____
_____

Exercise 21

Company name:	Exact location:
Person concerned:	Date:
Concrete pump blocking the access. What is missing?	Corrective action photo

Observation/findings:
_____
_____

Potential consequences/risks:
_____
_____

Recommended corrective action/Date:
_____
_____
_____
_____

Exercise 22

Company name:	Exact location:
Person concerned:	Date:
Full body harness on the ground—is this proper storage of fall arrest equipment?	Corrective action photo

Observation/findings:
_____
_____

Potential consequences/risks:
_____
_____

Recommended corrective action/Date:
_____
_____
_____

Exercise 23

Company name:	Exact location:
Person concerned:	Date:
Electrical crew working at this height. What needs to be done?	Corrective action photo

Observation/findings:
_____
_____

Potential consequences/risks:
_____
_____

Recommended corrective action/Date:
_____
_____
_____
_____

Exercise 24

Company name:	Exact location:
Person concerned:	Date:
What is missing on this drill press machine?	Corrective action photo

Observation/findings:
_____
_____

Potential consequences/risk:
_____
_____

Recommended corrective action/Date:
_____
_____
_____
_____

Exercise 25

Company name:	Exact location:
Person concerned:	Date:
What is missing on the hoist hook?	Corrective action photo

Observation/findings:
_____
_____

Potential consequences/risks:
_____
_____

Recommended corrective action/Date:
_____
_____
_____
_____

Exercise 26

Company name:	Exact location:
Person concerned:	Date:
Is this a safe location for the housekeeping tools?	Corrective action photo

Observation/findings:
_____
_____

Potential consequences/risk:
_____
_____

Recommended corrective action/Date:
_____
_____
_____
_____

Exercise 27

Company name:	Exact location:
Concern person:	Date:
What can you say about the stability of the scaffolding?	Corrective action photo

Observation/findings:
_____
_____

Potential consequences/risk:
_____
_____

Recommended corrective action/Date:
_____
_____
_____
_____

Exercise 28

Company name:	Exact location:
Person concerned:	Date:
Is this a safe working platform?	Corrective action photo

Observation/findings:
_____
_____

Potential consequences/risks:
_____
_____

Recommended corrective action/Date:
_____
_____
_____
_____

Exercise 29

Company name:	Exact location:
Person concerned:	Date:
Unsafe Condition	Corrective Action

Observation/findings:
_____
_____

Potential consequences/risks:
_____
_____

Recommended corrective action/Date:
_____
_____
_____
_____

Exercise 30

Company name:	Exact location:
Person concerned:	Date:
Unsafe Condition	Corrective Action

Observation/findings:
_____
_____

Potential consequences/risks:
_____
_____

Recommended corrective action/Date:
_____
_____
_____
_____

Exercise 32

Company name:	Exact location:
Person concerned:	Date:
Unsafe Condition	Corrective Action

Observation/findings:
_____
_____

Potential consequences/risks:
_____
_____

Recommended corrective action/Date:
_____
_____
_____
_____

Exercise 33

Company name:	Exact location:
Person concerned:	Date:
Unsafe emergency Exit	Corrective Action

Observation/findings:
_____
_____

Potential consequences/risks:
_____
_____

Recommended corrective action/Date:
_____
_____
_____
_____

Exercise 34

Company name:	Exact location:
Person concerned:	Date:
Unsafe Condition	Corrective Action

Observation/findings:
_____
_____

Potential consequences/risks:
_____
_____

Recommended corrective action/Date:
_____
_____
_____
_____

Exercise 35

Company name:	Exact location:
Person concerned:	Date:
Findings	Corrective Action

Observation/findings:
_____
_____

Potential consequences/risks:
_____
_____

Recommended corrective action/Date:
_____
_____
_____
_____

Exercise 36

Company name:	Exact location:
Person concerned:	Date:
Unsafe Condition	Corrective Action

Observation/findings:
_____
_____

Potential consequences/risks:
_____
_____

Recommended corrective action/Date:
_____
_____
_____
_____

Exercise 37

Company name:	Exact location:
Person concerned:	Date:
Unsafe Condition	Corrective Action

Observation/findings:
_____
_____

Potential consequences/risks:
_____
_____

Recommended corrective action/Date:
_____
_____
_____
_____

Exercise 38

Company name:	Exact location:
Person concerned:	Date:
Unsafe Condition	Corrective Action

Observation/findings:
_____
_____

Potential consequences/risks:
_____
_____

Recommended corrective action/Date:
_____
_____
_____
_____

Exercise 39

Company name:	Exact location:
Person concerned:	Date:
Unsafe Condition	Corrective Action

Observation/findings:
_____
_____

Potential consequences/risks:
_____
_____

Recommended corrective action/Date:
_____
_____
_____
_____

Exercise 40

Company name:	Exact location:
Person concerned:	Date:
Unsafe Condition	Corrective Action

Observation/findings:
_____
_____

Potential consequences/risks:
_____
_____

Recommended corrective action/Date:
_____
_____
_____

Exercise 41

Company name:	Exact location:
Person concerned:	Date:
Unsafe Condition	Corrective Action

Observation/findings:
_____
_____

Potential consequences/risks:
_____
_____

Recommended corrective action/Date:
_____
_____
_____
_____

Exercise 42

Company name:	Exact location:
Person concerned:	Date:
Unsafe Condition	Corrective Action

Observation/findings:
_____
_____

Potential consequences/risks:
_____
_____

Recommended corrective action/Date:
_____
_____
_____
_____

Exercise 43

Company name:	Exact location:
Person concerned:	Date:
Unsafe Condition	Corrective Action

Observation/findings:
_____
_____

Potential consequences/risks:
_____
_____

Recommended corrective action/Date:
_____
_____
_____
_____

Exercise 44

Company name:	Exact location:
Person concerned:	Date:
Unsafe Condition	Corrective Action

Observation/findings:
_____
_____

Potential consequences/risks:
_____
_____

Recommended corrective action/Date:
_____
_____
_____
_____

Exercise 45

Company name:	Exact location:
Person concerned:	Date:
What is required PPE for this job?	Corrective action photo

Observation/findings:
_____
_____

Potential consequences/risk:
_____
_____

Recommended corrective action/Date:
_____
_____
_____
_____

Exercise 46

Company name:	Exact location:
Person concerned:	Date:
Unsafe Condition	Corrective Action

Observation/findings:
_____
_____

Potential consequences/risks:
_____
_____

Recommended corrective action/Date:
_____
_____
_____
_____

Exercise 47

Company name:	Exact location:
Person concerned:	Date:
What is required PPE for this job?	Corrective Action

Observation/findings:
_____
_____

Potential consequences/risks:
_____
_____

Recommended corrective action/Date:
_____
_____
_____
_____

Exercise 48

Company name:	Exact location:
Person concerned:	Date:
Unsafe Condition	Corrective Action

Observation/findings:
_____
_____

Potential consequences/risks:
_____
_____

Recommended corrective action/Date:
_____
_____
_____
_____

Exercise 49

Company name:	Exact location:
Person concerned:	Date:
Unsafe Condition	Corrective Action

Observation/findings:
_____
_____

Potential consequences/risks:
_____
_____

Recommended corrective action/Date:
_____
_____
_____
_____

Exercise 50

Company name:	Exact location:
Person concerned:	Date: |Н

What is the missing part of the hoist hook?	Corrective Action

Observation/findings:
_____
_____

Potential consequences/risks:
_____
_____

Recommended corrective action/Date:
_____
_____
_____
_____

Exercise 51

Company name:	Exact location:
Person concerned:	Date:
Unsafe Condition	Corrective Action

Observation/findings:
_____
_____

Potential consequences/risks:
_____
_____

Recommended corrective action/Date:
_____
_____
_____
_____

Exercise 52

Company name:	Exact location:
Person concerned:	Date:
Unsafe Condition	Safe Condition

Observation/findings:
_____
_____

Potential consequences/risks:
_____
_____

Recommended corrective action/Date:
_____
_____
_____
_____

Exercise 53

Company name:	Exact location:
Person concerned:	Date:

Unsafe Condition	Corrective Action

Observation/findings:
_____
_____

Potential consequences/risks:
_____
_____

Recommended corrective action/Date:
_____
_____
_____
_____

Exercise 54

Company name:	Exact location:
Person concerned:	Date:
Unsafe Condition	Corrective Action
Observation/findings:	
Potential consequences/risks:	
Recommended corrective action/control:	

Exercise 55

Company name:	Exact location:
Person concerned:	Date:
Unsafe Condition	Corrective action photo

Observation/findings:
_____
_____

Potential consequences/risks:
_____
_____

Recommended corrective action/Date:
_____
_____
_____
_____

Exercise 56

Company name:	Exact location:
Person concerned:	Date:
Is this safety condition?	Corrective Action

Observation/findings:
_____
_____

Potential consequences/risk:
_____
_____

Recommended corrective action/control:
_____
_____
_____
_____

Exercise 57

Company name:	Exact location:
Person concerned:	Date:
Look at the Fire Extinguisher pressure gauge? #13	Corrective Action

Observation/findings:
_____
_____

Potential consequences/risks:
_____
_____

Recommended corrective action/Date:
_____
_____
_____
_____

Exercise 58

Company name:	Exact location:
Person concerned:	Date:
Unsafe Act	Corrective Action

Observation/findings:
_____
_____

Potential consequences/risks:
_____
_____

Recommended corrective action/Date:
_____
_____
_____
_____

Exercise 59

Company name:	Exact location:
Person concerned:	Date:
Unsafe Condition — EQUIPMENT RENTAL 03 857 6769	Corrective Action

Observation/findings:
_____
_____

Potential consequences/risks:
_____
_____

Recommended corrective action/Date:
_____
_____
_____
_____

Exercise 60

Company name:	Exact location:
Person concerned:	Date:
Unsafe Condition — Can you see the wrong condition?	Corrective Action

Observation/findings:
_____
_____

Potential consequences/risks:
_____
_____

Recommended corrective action/Date:
_____
_____
_____
_____

Exercise 61

Company name:	Exact location:
Person concerned:	Date:
Unsafe Condition	Corrective Action

Observation/findings:
_____
_____

Potential consequences/risks:
_____
_____

Recommended corrective action/Date:
_____
_____
_____
_____

Exercise 62

Company name:	Exact location:
Person concerned:	Date:
Is this safe condition?	Corrective Action

Observation/findings:
_____
_____

Potential consequences/risks:
_____
_____

Recommended corrective action/Date:
_____
_____
_____
_____

Exercise 63

Company name:	Exact location:
Person concerned:	Date: |</br>*Unsafe act – is this the right way to carry a power tool?*	Corrective Action

Observation/findings:
_____
_____

Potential consequences/risks:
_____
_____

Recommended corrective action/Date:
_____
_____
_____
_____

Exercise 64

Company name:	Exact location:
Person concerned:	Date:
Unsafe Act	Corrective Action

Observation/findings:
_____
_____

Potential consequences/risks:
_____
_____

Recommended corrective action/Date:
_____
_____
_____
_____

Exercise 65

Company name:	Exact location:
Person concerned:	Date:
Unsafe Act	Corrective Action

Observation/findings:
_____
_____

Potential consequences/risks:
_____
_____

Recommended corrective action/Date:
_____
_____
_____
_____

Exercise 67

Company name:	Exact location:
Person concerned:	Date:
Unsafe Condition	Corrective Action

Observation/findings:
_____
_____

Potential consequences/risks:
_____
_____

Recommended corrective action/Date:
_____
_____
_____
_____

Exercise 68

Company name:	Exact location:
Person concerned:	Date:
Findings	Corrective Action

Observation/findings:
_____
_____

Potential consequences/risks:
_____
_____

Recommended corrective action/Date:
_____
_____
_____
_____

Exercise 69

Company name:	Exact location:
Person concerned:	Date:
Unsafe Act/Condition	Corrective Action

Observation/findings:
_____
_____

Potential consequences/risks:
_____
_____

Recommended corrective action/Date:
_____
_____
_____
_____

Exercise 70

Company name:	Exact location:
Person concerned:	Date:
Unsafe Condition — bus tire	Corrective Action

Observation/findings:
_____
_____

Potential consequences/risks:
_____
_____

Recommended corrective action/Date:
_____
_____
_____
_____

Conducting site inspection

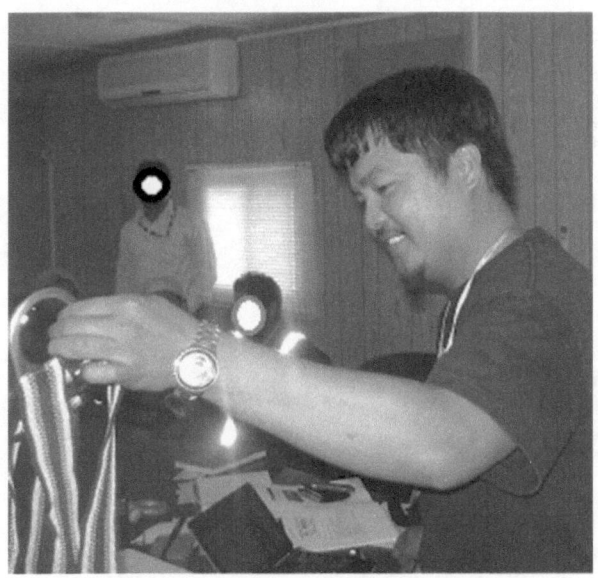

Conducting fall protection training

Confined space activity safety requirements verification

Monitoring Critical Lift Operation

Completed fire extinguisher training

Coaching session with mentors

A Message of Thanks from the Author

"Alhamdulillah" Thanks and Praise to Allah our Almighty Creator for giving me the wisdom to create this simple handbook; I dedicate this handbook to my co- safety practitioners and to those who are willing to join safety profession in the future.

Thank you to my family who has been my inspiration in completing this book. Indeed, these will not be a success without your love and support.

To my PSSP-MER couleagues and brother, friends, all safety organizations, to my employers, supervisors and mentors thanks you for all your supports. I owe you my success in this profession.

To all my APO brothers may we always be LFS.

"Assalamu Alaikum Wa Rahmatullahi Wa Barakatu". Jazzak Allah Kayran.

www.ingramcontent.com/pod-product-compliance
Lightning Source LLC
Chambersburg PA
CBHW030801180526
45163CB00003B/1115